Chemistry and the human genome

Ted Lister

Written by Ted Lister

Edited by Colin Osborne and Maria Pack

Designed by Oxford Designers & Illustrators Ltd.

Published and distributed by Royal Society of Chemistry

Printed by Royal Society of Chemistry

Copyright © Royal Society of Chemistry 2002

Registered charity No. 207890

For further information on other educational activities undertaken by the Royal Society of Chemistry contact: Email : education@rsc.org
Tel   : 020 7440 3344

Education Department
Royal Society of Chemistry
Burlington House
Piccadilly
London W1J 0BA

Information on other Royal Society of Chemistry activities can be found on its websites:
www.rsc.org
www.chemsoc.org
www.chemsoc.org/LearnNet contains resources for teachers and students from around the world.

ISBN 0-85404-396-9

British Library Cataloguing in Publication Data.

A catalogue for this book is available from the British Library.

Contents

Acknowledgements

The author and the Royal Society of Chemistry thank the following for their help in preparing this material:

Georgina Day, Institute of Biology;

Jon Evans, Royal Society of Chemistry;

Carole Foy, The Laboratory of the Government Chemist;

Melanie Wall, The Wellcome Trust Sanger Institute; and

Professor Ed Wood, School of Biochemistry and Molecular Biology, University of Leeds, Chairman of the Education Group of the Biochemical Society.

Foreword

The discovery of the sequence of the human genome has been hailed as one of the most important scientific advances ever. This would not have been possible without the work of chemists and chemical scientists. This resource aims to show students and teachers the chemical science behind the discovery and to place it in the context of science in the 21st century.

It is hoped that teachers of both biology and chemistry at post-16 level will use it to enthuse their students to perhaps go on to study the proteome and beyond.

Professor Steven Ley CChem FRSC FRS
President, Royal Society of Chemistry

How to use this book

The publication of the draft sequence of human genome is undoubtedly one of mankind's greatest scientific achievements. This book offers an account of the chemical science that underlies this achievement at a level suitable for post-16 students of chemistry and also biology. Teachers of both chemistry and biology may wish to use it to update their own knowledge and the book may also be used directly by students. To this end questions have been included within the text (with answers for the teacher at the back of the book). Key words in bold in the text are explained in a glossary. Some suggestions for further reading are included as well as a short list of web sites (although many more of these can be found by using a search engine).

Some ways in which the book may be used are listed:

- It could be used by students as background reading to inspire them and give them a feel for the context of this particular topic.
- Sections could be used together with the questions they contain as comprehension exercises or extracts could be used as passages on which to base comprehension questions for exam practice.
- The reading of a section and the answering of the questions within it could be set as a meaningful exercise during the absence (planned or unplanned) of the teacher.
- Students could read the book simply for interest and enjoyment using the questions as an aid to understanding as they read.
- Teachers themselves might read the book for interest and enjoyment as a means of updating their own knowledge.

1 Introduction

On 15th February 2001, in two articles in the magazines *Nature* and *Science*, two groups of scientists published a draft that described the complete chemical makeup of human genes – the human **genome** as it is often called. Many people have referred to this as the greatest scientific achievement of the 20th century for, although the announcement was made in the 21st century (and work on the details will continue for the foreseeable future), the project can be traced back to the 1980s. In effect, what the scientists have found is the chemical 'code' or 'recipe' for making a human being.

To look at, this recipe would not be very interesting. If it were printed out, it would consist simply of a sequence of the four letters A, C, G and T that would fill the equivalent of 200 telephone directories and take nine years to read aloud. However, the significance of this achievement is certain to be enormous.

Reprinted with permission from *Nature* 409: 15th February 2001. Copyright 2001 Macmillan Publishers Ltd.

Reprinted with permission from *Science* 16th Feb 2001 Vol 291 No. 5507. Copyright 2001 American Association for the Advancement of Science.

Figure 1 *The draft of the human genome was published simultaneously in* Nature *and* Science

Scientific journals – how scientists share information

Scientists in universities and other publicly funded institutions (such as hospitals, museums *etc*) have traditionally made their results freely available to others. Such scientists are sometimes called academic scientists. They share their results by publishing them in scientific journals – magazines that come out regularly – that other scientists read. Increasingly, these are appearing in electronic as well as paper-based versions. This is how scientific information is passed on and it is how academic scientists make their reputations. There is a wide range of different journals, some, such as *Nature*, deal with the whole of science while others are narrower in their focus, dealing with organic chemistry or reaction kinetics, for example. Before an article, usually called a paper, is published, it is read by other respected scientists working in the same field who ensure that it is original, that it represents good science and that it is worthy of publication. This process is called **peer review**. Institutions where scientists work will subscribe to many of these journals. The Royal Society of Chemistry publishes a number of the most prestigious journals in the field of chemistry, for example.

Researchers in private companies (such as pharmaceutical companies) have always tended to patent their discoveries. Patents establish ownership of a discovery or invention and allow the patent holder to make money from it. This is justified on the grounds that research is expensive and that companies must make money from their discoveries in order to finance more research in the future, as well as to make a profit for their shareholders.

Scientists working for the military have, for obvious reasons, tended to keep much of their work secret.

Scientists in both private companies and the military do, of course, have access to all the work published by academic scientists.

Private *versus* public – different approaches to the genome

There are two main organisations working on the sequencing of the human genome – The International Human Genome Sequencing Consortium often called The Human Genome Project (HGP) and Celera Genomics. HGP is funded by public finance (governments and charities) and includes institutions in many countries working together, while Celera is private company. HGP is committed to making public, without charge, the information it discovers. It does this *via* the internet, and, each day, new base sequences are placed on its website where they can be freely accessed by anyone (www.sanger.ac.uk/, accessed May 2002). Celera places severe restrictions on others using its data while it itself, of course, has access to HGP's information and uses it in its publications. Celera hopes to make money by discovering and patenting gene sequences that might prove to be useful – in medical treatments for example.

There has been considerable debate over this public-private split on a number of grounds. Some of the points that have been made are as follows:

● Is it morally right to patent basic scientific knowledge? Some people have drawn a parallel with patenting the Periodic Table.

● What is the legality of patenting genetic information? Who actually owns the information – the person whose genes are being investigated or the scientist that decodes the information or the whole human race?

● Unless genetic information can be patented, there will be no incentive for commercial firms to pay for research. Celera's supporters argue that the incentive of patents will attract private investors to put money into scientific research, which they would not do if the fruits of that research were made freely available to all.

● If some information is patented, this may prevent other scientists making important discoveries because they cannot freely use the patented data.

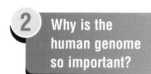

Why is the human genome so important?

We have always known that children resemble their parents in various ways and, perhaps more importantly, that humans produce human offspring, not chimpanzees, houseflies or tulips, and that chimpanzee offspring are chimpanzees and so on. This is called heredity – we inherit characteristics from our parents. The science of studying inheritance is called genetics. Finding out how inheritance happens has been a puzzle that scientists have worked on for centuries (see Box *Why do we resemble our parents?*) and the map of the human genome is the latest step of this process.

This knowledge is important in two different ways. Firstly, scientists want to understand the world around them, how it works and why it is as it is; so unravelling the human genome is a major intellectual achievement. Secondly there will be enormous practical spin-offs from this understanding. For example many diseases, such as muscular dystrophy, cystic fibrosis and sickle cell anaemia, are inherited, or have a hereditary component. Understanding the genome brings closer the possibility of testing for these diseases and ultimately being able to treat them. This testing might be done on an individual with a disease to confirm that its cause was genetic or on tissue from a foetus in the womb to confirm that it had no genetic disease. More controversially it also brings closer the possibility of cloning (making exact copies of) human beings.

Why do we resemble our parents?

From realising that children resemble their parents to sequencing the human genome has been a long journey. Here are some of the milestones along the road.

1665	Robert Hooke describes cells
1831	Robert Brown describes the cell nucleus
1836	Hugo von Mohl describes cell division
1865	Gregor Mendel publishes his work on breeding pea plants
1869	Johann Frederick Miescher identifies nucleic acid in cells
1875	Eduard Strasburger describes chromosomes
1885	Albrecht Kossel identifies the sugar and the bases in nucleic acids
1948	Erwin Chargaff notices that in nucleic acids the amount of guanine (G) is equal to the amount of cytosine (C) and that the amount of adenine (A) is equal to the amount of thymine (T)
1953	Francis Crick, Rosalind Franklin, James Watson and Maurice Wilkins discover the structure of DNA
1966	The genetic code deciphered
1977	Fred Sanger publishes his method for DNA sequencing
1990	The Human Genome Project is set up
1996	First human gene map published
1998	Celera is set up
1999	First human chromosome completely sequenced
2001	Draft of the human genome published

3 How does heredity work?

In the 1860s, an Austrian monk called Gregor Mendel worked out the basic rules of heredity for a particularly simple case. He found, for example, that if you cross a tall pea plant with a dwarf one, three-quarters of the offspring are tall, and one-quarter dwarf. However, Mendel had no idea why this should be so – how the offspring 'knew' whether they should be tall or dwarf, *ie* how the information was carried from one generation to the next.

It was known, however, that all living organisms are made of cells and that when organisms reproduce, cells divide. Shortly after Mendel's discovery it was noticed that during cell division, structures in the nuclei of cells called **chromosomes** divide into two, one passing into each of the two new cells, see Figure 2. This suggested that chromosomes might carry genetic information. Later it was found that chromosomes consist of proteins and a material called **deoxyribonucleic acid (DNA)** in approximately equal amounts. At first, the protein was suspected of being the information-carrier because of the importance of proteins elsewhere in biochemistry but eventually it was shown to be the DNA. Originally DNA was thought to be too simple in structure to carry much information – the idea of digital information had not really surfaced at the time. So heredity is due to the transfer of DNA from parents to their offspring.

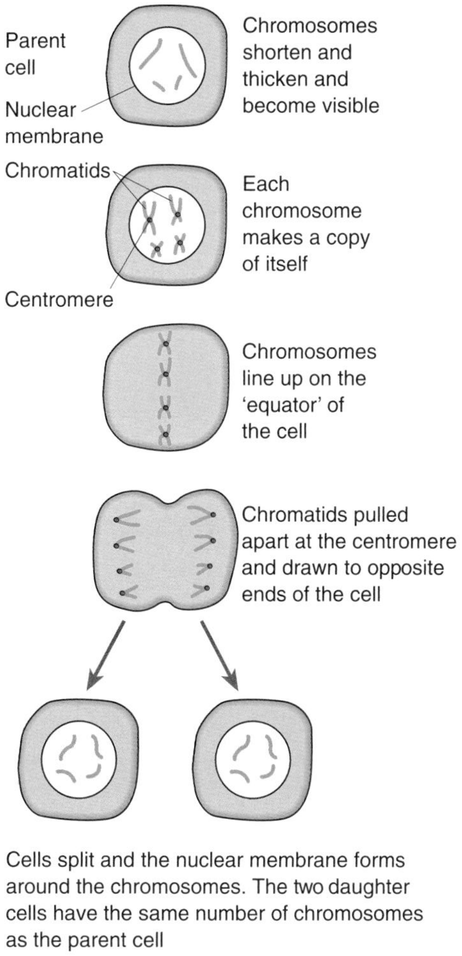

Parent cell — Chromosomes shorten and thicken and become visible

Nuclear membrane

Chromatids — Each chromosome makes a copy of itself

Centromere

Chromosomes line up on the 'equator' of the cell

Chromatids pulled apart at the centromere and drawn to opposite ends of the cell

Cells split and the nuclear membrane forms around the chromosomes. The two daughter cells have the same number of chromosomes as the parent cell

Figure 2   *Cell division*
Adapted with permission from *Access to A-level biology,* H. Bowen, J. Good; Stanley Thornes, 1996.

In the 1940s and 50s, chemists worked out the structure of DNA and it turned out to be a polymer made from just four different types of monomers called **nucleotides**. Each nucleotide consists of three parts – a sugar called **deoxyribose** (strictly 3-deoxyribose), a **phosphate group** and an organic **base**, see Figure 3.

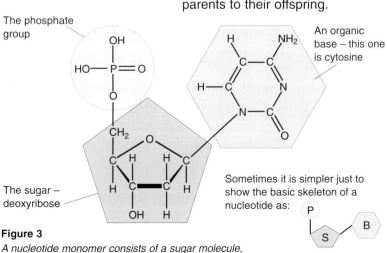

The phosphate group

An organic base – this one is cytosine

The sugar – deoxyribose

Sometimes it is simpler just to show the basic skeleton of a nucleotide as:

Figure 3
A nucleotide monomer consists of a sugar molecule, a phosphate group and an organic base. See the section 'Representation of phosphate, sugar and base units' in the teachers' notes for information on the conventions used to represent these in diagrams.

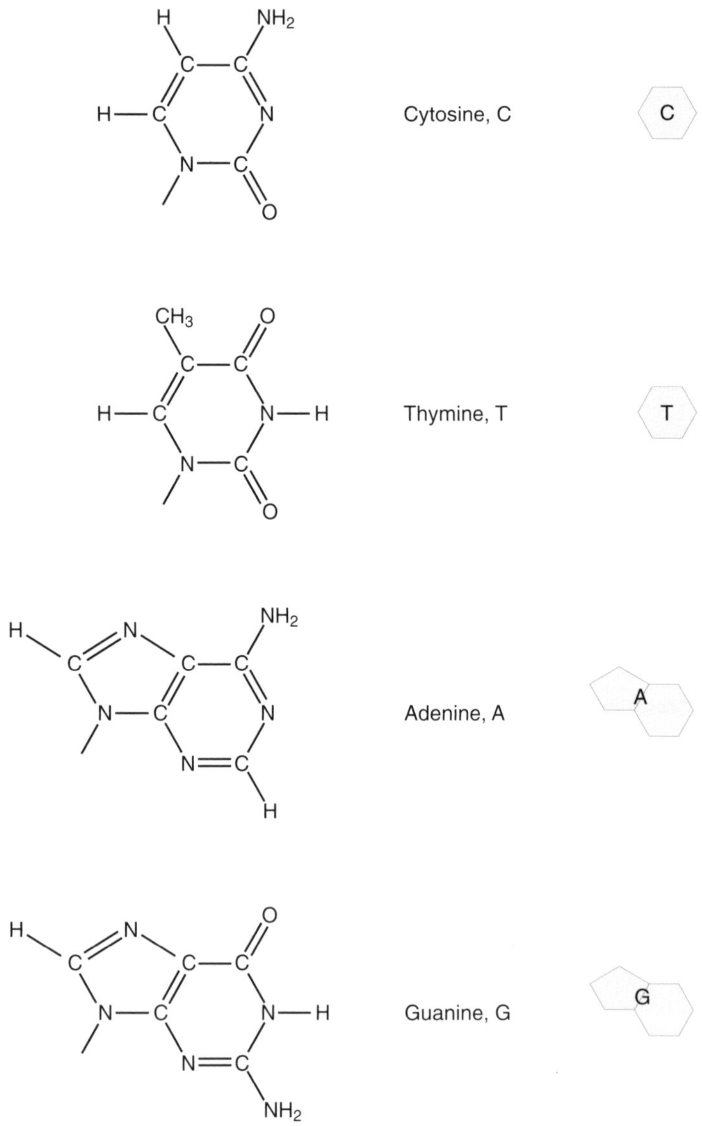

The sugar and the phosphate are the same in all nucleotide molecules but there are four different bases, called adenine, cytosine, guanine, and thymine, usually abbreviated to A, C, G and T, see Figure 4.

Q 1. Adenine is the most basic of the four bases found in DNA. What feature of its molecule makes it basic?

Q 2. What feature of the DNA molecule makes it acidic (see Figure 6)?

Two nucleotides can react together, an –OH group of a phosphate on one nucleotide reacting with an –OH group on the sugar of the other to eliminate a molecule of water, see Figure 6. (p. 11)

More nucleotide molecules can add on in the same way to form a polymer chain called DNA. If there were only one type of base, there would only be one type of DNA and the only variation would be the number of nucleotides in the chain, the **chain length**. However, because there are four different bases, there is an infinite variety of possible DNA polymers which differ in the sequence of bases along the chain. These sequences can be written using the abbreviations for the bases to show the arrangement, *eg* ATGCCAATTTG *etc.* In effect the sequence of bases holds information digitally rather like data is stored on a computer in binary notation – a sequence of 1s and 0s. Figure 5 shows a very short section of a single strand of a DNA molecule.

A single DNA molecule may contain many millions of these monomers arranged in any order. Scientists measure the length of the chain in units called kilobases (kb, thousands of bases) or megabases (Mb, millions of bases) or even gigabases (Gb, thousand millions of bases).

Cytosine, C   C

Thymine, T   T

Adenine, A   A

Guanine, G   G

Figure 4
The four organic base units found in DNA

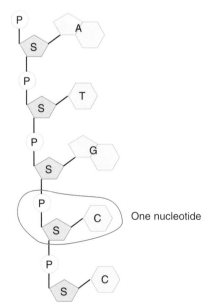

Figure 5
A section of a single strand of a DNA molecule

One nucleotide

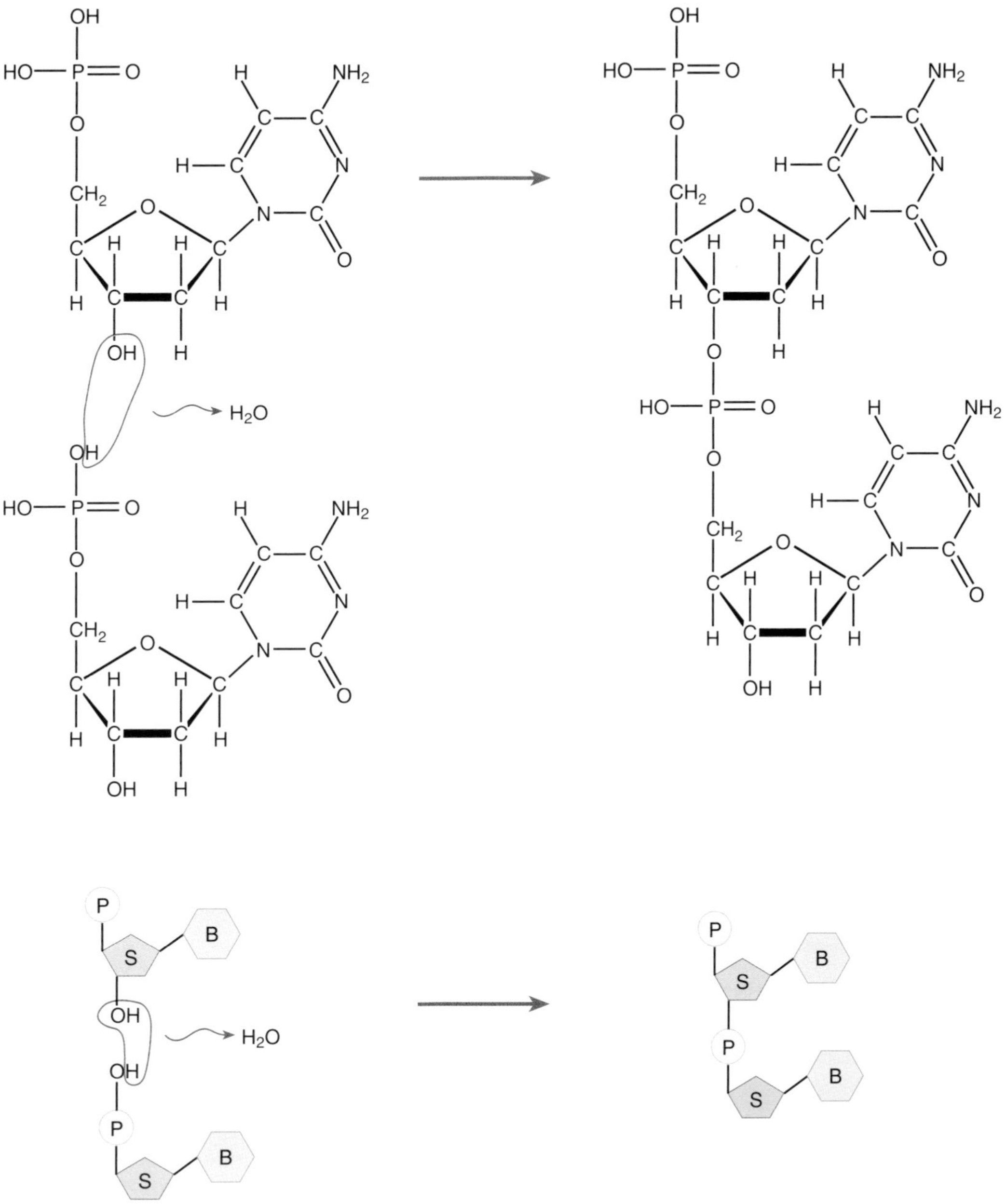

Figure 6 *Two nucleotides can react together*

This, of course means that there are vast numbers of possible DNA molecules each with a different order of nucleotides along the chain. The order of the nucleotides on the chain is what holds the information used to make a human being (or other living creature). This is like saying that in a recipe for a cake it is the order of the 26 letters of the alphabet that holds the information by forming the words of the recipe.

The bases form a simple four-letter alphabet. The different bases correspond to letters, and sequences of bases mean something, just like sequences of letters make meaningful words, sentences, paragraphs, chapters and books. So, just as we can read the recipe for making a cake from the sequence of letters in a cookery book, cells of the human body can 'read' the instructions for making a human being from the sequence of nucleotides in a DNA molecule. It is this sequence of bases that the scientists unravelling the human genome have 'read'. In humans, the DNA is divided among 23 pairs of structures called chromosomes found in the nucleus of each cell. Each chromosome contains a molecule of DNA (strictly speaking a pair of molecules – see below) that is tightly coiled. The coiling is so tight that the DNA in a cell has a diameter of about 6 μm (6 millionths of a metre) but would be about 2 m long if uncoiled.

This is fine as far as it goes, but it leads to three further questions:

1. When cells divide, both of the new cells have the genetic information; how is the information copied when cells divide?
2. How is the 'recipe' turned into a human being, or other organism? and
3. Why are offspring not *exactly* like one of their parents?

How is DNA copied?

The answer to this question was found in the 1950s by two groups of young scientists – Francis Crick and James Watson working in Cambridge, and Rosalind Franklin and Maurice Wilkins in London, Figure 7. Franklin and Wilkins used a technique called X-ray diffraction to investigate DNA. Nowadays, with technical advances and the help of fast computers, X-ray diffraction can be used to locate the exact positions of atoms in complex structures quite quickly. In the 1950s, it was a painstakingly slow process and could do no more than give a clue to the structure of DNA. This vital clue was that the molecule's chain formed a spiral, or helix, with one complete turn of the spiral every 3.4 nm.

The Double Helix – finding the structure of DNA

Francis Crick

James Watson

Maurice Wilkins

Rosalind Franklin

Figure 7

Most people associate the names of Francis Crick and James Watson with the discovery of the double helix structure of DNA. They shared their Nobel Prize for this feat with Maurice Wilkins who initially suggested that the structure might be a spiral based on the evidence of X-ray diffraction. However, a fourth contributor to this discovery is less well known. This is Rosalind Franklin who actually obtained the X-ray diffraction patterns. The four could not be described as a team and were often at loggerheads. Franklin's contribution has often been underplayed – partly, perhaps, because of anti-feminist attitudes prevalent in the 1950s scientific world. However, Franklin's failure to share in the prize was not due to this; she died in 1958, at the age of 37, four years before the Nobel Prize was awarded. Nobel Prizes are never awarded posthumously.

James Watson wrote a book, *The Double Helix*, (see *Further reading*) about the discovery of the DNA structure. It is well worth reading. It gives an excellent account of the science involved (without too much technical detail) and gives a fascinating insight into the personal relationships and lifestyles of the people involved.

Crick and Watson worked with molecular models. They used crude cardboard ones at first and then precision-made ones specially constructed by the Cambridge University workshops – off-the-shelf plastic models were not available at that time. X-ray diffraction showed that DNA was in fact a *double* helix – two intertwined spirals. Crick and Watson showed that these two spirals consisted of a 'backbone' made of the linked sugar and phosphate molecules with the bases on the inside of the spiral. The two spirals were held together by **hydrogen bonds** (see box *Hydrogen bonding*) between the bases. The base adenine (A) can form two hydrogen bonds with thymine (T) while guanine (G) and cytosine (C) can form three hydrogen bonds with each other, see Figure 8. This is called **base pairing**. The shapes of the molecules mean that other pairs, such as T and C hydrogen bond together much less effectively than A and T and G and C.

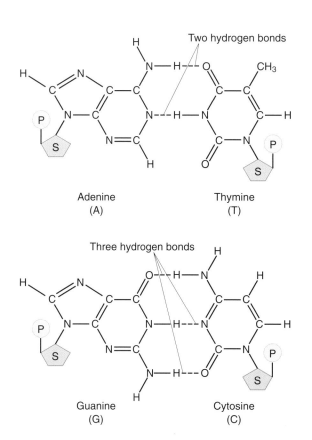

Figure 8 *Base pairing. A hydrogen bond is normally represented by three dashes --- . A can pair effectively only with T and vice versa. G can pair effectively only with C and vice versa.*

This image gives an impression of the double helix structure of DNA.

Q 3. Make a model of a single nucleotide using a commercial molecular modelling kit. You may make it using any of the four bases. How long does it take? Imagine how long it would take using cardboard and Sellotape™ to make several nucleotides linked together. Now work with another group of students that has also made a model and show how the two molecules can link together via a sugar-phosphate bond. Use Figure 6 as a guide. What small molecule is eliminated in the process? What is this type of reaction called? Keep your model for the next question.

Hydrogen bonding

There are three types of intermolecular forces – hydrogen bonding, dipole-dipole bonding and van der Waals interactions. Hydrogen bonding is the strongest of these. It occurs only in very specific circumstances – between a hydrogen atom bonded covalently to an atom of fluorine, oxygen or nitrogen and another atom of fluorine, oxygen or nitrogen (although in biological systems, we can ignore fluorine because it is relatively rarely found). The hydrogen atom becomes sandwiched between the two other atoms and holds them together.

The bonding works like this. Oxygen and nitrogen are strongly **electronegative** atoms, that is, when they form covalent bonds with other atoms, they draw the electrons in the bond towards themselves. So in an O–H or N–H bond, the oxygen or nitrogen has a partial negative charge and the hydrogen a partial positive charge. They are written $O^{\delta-}$ (or $N^{\delta-}$) and $H^{\delta+}$. So the hydrogen atom will attract another oxygen (or nitrogen) electrostatically, Figure 9.

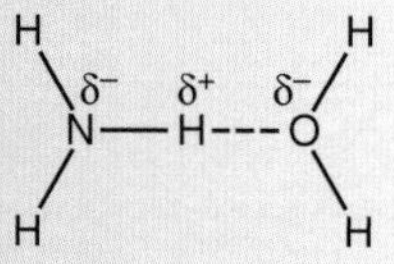

Figure 9 *A hydrogen bond (---) between a molecule of ammonia and one of water*

However, there is more to hydrogen bonding than this. Since a hydrogen atom has just one electron and the electronegative atom has partially removed it, a nearby atom will 'see' what is in effect a 'bare' hydrogen nucleus – a proton. Since a proton is so small, it has an extremely strong electric field and will powerfully attract lone pairs of electrons on electronegative atoms towards it. Thus it forms something very close to a second covalent bond between the hydrogen atom and the other electronegative atom, Figure 10.

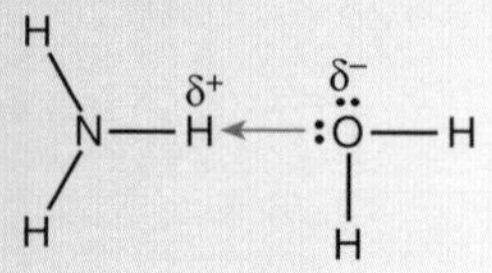

Figure 10 *A lone pair of electrons form the hydrogen bond between water and ammonia*

Crick and Watson's molecular models showed that the shape of the molecules and angles of the bonds fit the double helix pattern perfectly and that the helix repeated itself every 3.4 nm as the X-ray evidence showed. They also showed that, because of the shapes of the base molecules, cytosine can hydrogen bond most effectively with guanine, and thymine with adenine. This 'base pairing' means that the two strands of DNA in the double helix are different – where one has C, the other has G and where one has T, the other has A. The two strands are said to be **complementary**. However, they both contain the same information because the sequence of bases on one strand can easily be worked out from that on the other simply by replacing each base with its complement – C for G, G for C, T for A and A for T, Figure 11.

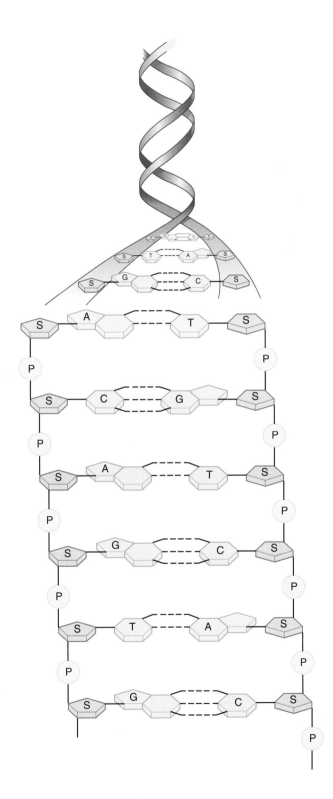

Figure 11 *The double helix of DNA is held together by hydrogen bonds between the paired bases*

Q 4. Imagine you have a short strand of DNA whose base sequence is AGGTCAAT. What is the base sequence of its complementary strand (reading from the same end)?

Q 5. Take your molecular model of a nucleotide from Q3 and find someone else in the group who has made a complementary nucleotide. See how the two molecules fit together by hydrogen bonding between the bases using Figure 8 as a guide.

The fact that the spirals are held together with hydrogen bonds is vital in explaining the copying process. Hydrogen bonds have roughly 5–10% of the strength of covalent bonds. This means that at room temperature, the hydrogen bonds holding the two spirals together can break (and re-form) while the covalent bonds holding the chains of the spiral together cannot.

Q 6. What are the strengths of a typical covalent bond, such as C-C and a typical hydrogen bond such as H---O in water? Give an approximate figure first then look up accurate figures in a data book or database.

In the nucleus of the cell, DNA double helix molecules exist along with a 'soup' of separate nucleotide monomers of all four types. (In fact this is an oversimplification – see Box *Reactions between nucleotides* – but it does not affect the essential argument.) If a double helix of DNA partially unravels due to the hydrogen bonds between the bases breaking, free nucleotide molecules of the correct type may 'move in' and their bases pair up with the newly exposed bases on the two unravelled strands (C pairing with G and A with T *etc*). These nucleotides, held in place by hydrogen bonding between the bases, then link together by phosphate-sugar bonding. The linking process is catalysed by an enzyme called DNA polymerase. In this way the original strand of DNA acts as a template for a new one, Figure 12.

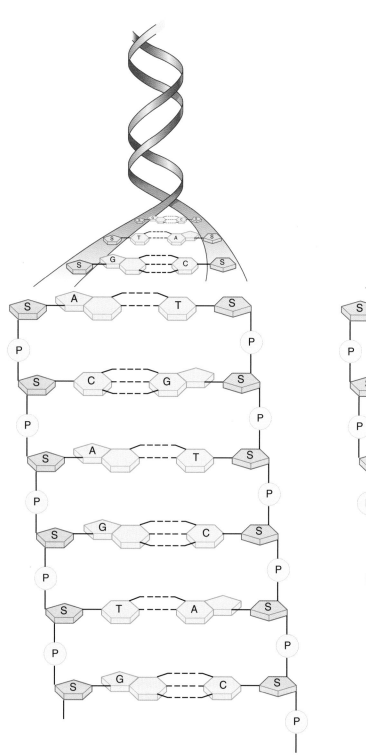

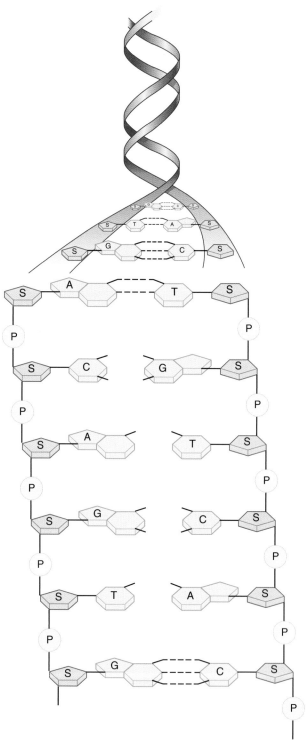

Figure 12 *The original DNA ...*   *... partially unravels ...*

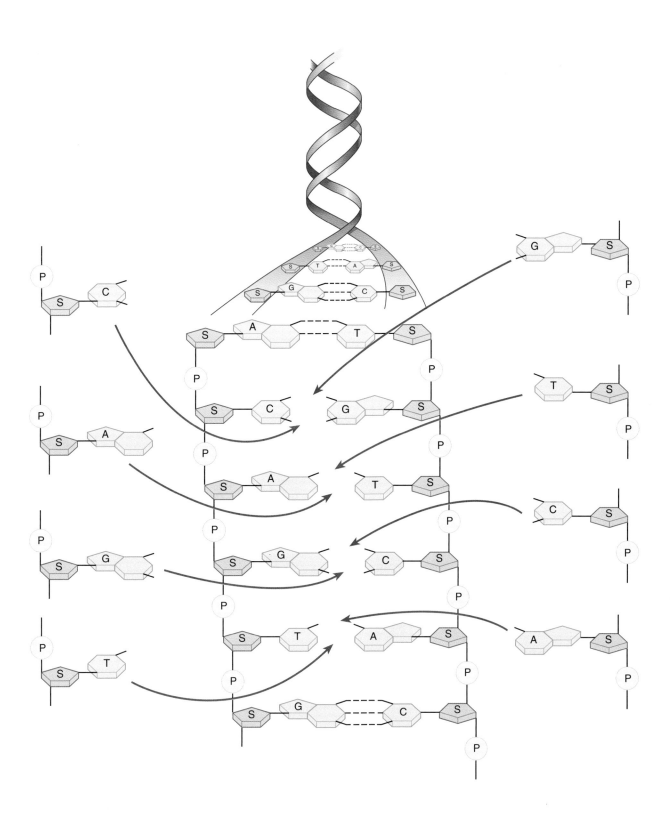

Figure 12 (cont'd) *... free nucleotides pair up with their complementary bases on each strand ...*

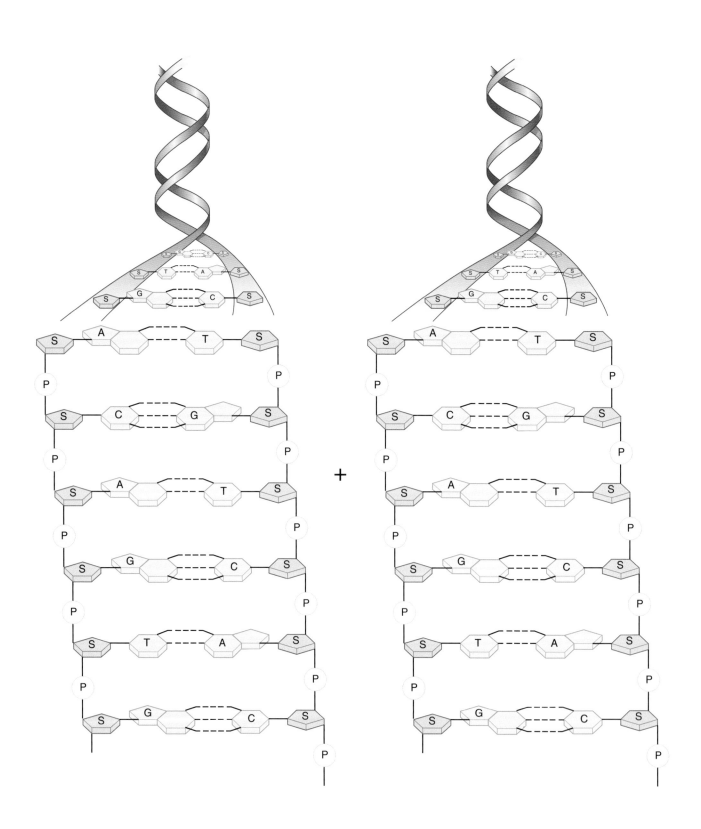

Figure 12 (cont'd) *... and two new strands identical to the original are formed*

The ultimate result of this is the formation of two identical double helices, each identical to the first. The sequence of nucleotides has in effect been copied. This process, called **replication**, occurs when chromosomes divide – one double helix passing into one new cell and the other into the other. So the instructions for making a human are passed on to the two new cells.

Reactions between nucleotides

The treatment of replication in the main text is somewhat oversimplified and this box describes it in more detail. While two nucleotides *can* react together as shown in Figure 6, this does not in fact occur in living systems. The species that actually react are **nucleoside** triphosphates, Figure 13, rather than monophosphates. (Take care with the two similar sounding words nucleotide and nucleoside – a nucleotide consists of a sugar linked to an organic base and a single phosphate group, a nucleoside consists of a sugar linked to an organic base only.) This is because the synthesis of large molecules such as DNA is endothermic – it requires an input of energy. This energy is supplied by the process of bond breaking and bond making during hydrolysis (reaction with water) of the triphosphate and is followed by loss of a diphosphate ion.

In more detail, what happens is as follows. The double helix of the existing DNA partially unwinds as a result of the breaking of the hydrogen bonds between the bases (enzymes help to control this process so that the whole double helix does not completely unwind). Then the nucleoside triphosphates move in and their bases pair up by hydrogen bonding with the newly exposed bases on the two unravelled strands (C pairing with G, and A with T). Hydrolysis of the triphosphates then takes place followed by the formation of a new sugar-phosphate link. This is how a base-sugar-phosphate unit is added to the new growing DNA molecule. In this way each original strand of DNA acts as a template for the new one.

The end result of this is the formation of two identical double helices, each identical to the first. Each one of these has one 'old' strand of DNA and one 'new' strand, and the sequence of bases has been faithfully copied. This process, called replication, occurs when cells divide, one double helix passing into one new cell and the other into the other. Each chromosome is one very long DNA double helix and it is important that the copying process is highly precise. There are 'proof reading' mechanisms to ensure this – the enzyme that controls DNA polymerisation can detect if a wrong base has been inserted, remove it and put in the correct one. In this way the instructions for making a human are passed on to the two new cells.

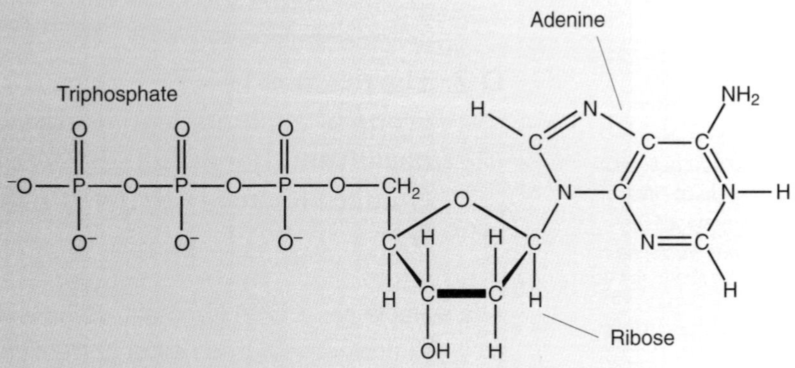

Figure 13 *A nucleoside triphosphate*

In order to 'read' the information carried by DNA molecules, we need to know the sequence of bases along the chain just as we need to know the sequence of letters in order to read a word or a sentence.

The method used for sequencing DNA involves breaking a chain of DNA, possibly many thousands of bases long, into short overlapping sections each consisting of perhaps 500 bases. These sections can be separated according to the number of bases they contain (essentially by their size) by the method of **gel electrophoresis** and other developments of this technique that use the same principles. Each section then has its sequence of bases determined by a technique called the **dideoxy method**. Powerful computers are then used to reassemble the overlapping sections so that the whole base sequence can be found.

DNA can be broken into sections using enzymes or by **sonication**, a method that uses ultrasound. Enzymes break the DNA molecule at specific points while sonication breaks it randomly.

Gel electrophoresis

This technique can be used to separate fragments of DNA according to their size (*ie* length). It relies on the fact that all fragments of DNA have an electric charge. This is because in body fluids, the free –OH group of the phosphate group of each nucleotide loses a H^+ ion, leaving behind a negatively charged ion, see Figure 14.

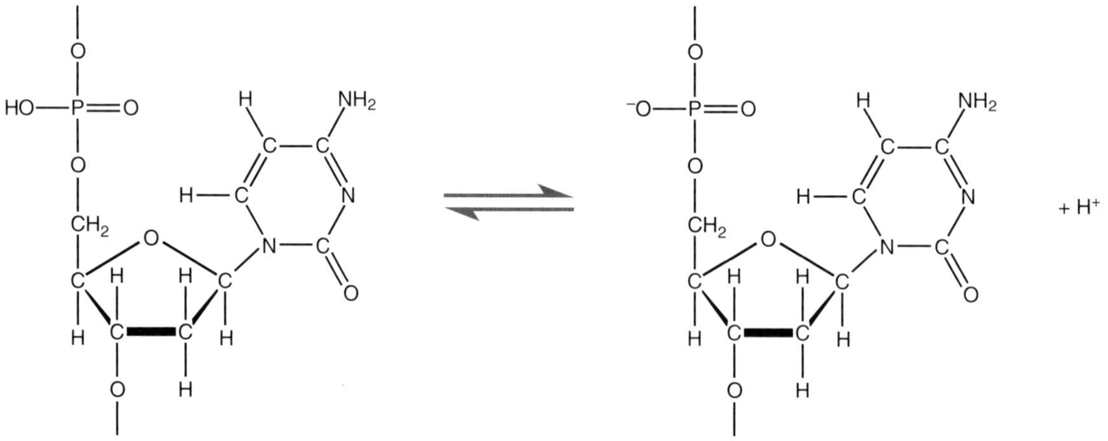

Figure 14

In aqueous solution the phosphate group of each nucleotide loses an H^+ ion leaving behind a negative ion. So each fragment of DNA carries as many negative charges as it has nucleotides

Q 7. The reaction by which the free –OH group of the phosphate group of each nucleotide loses a H^+ ion, leaving behind a negatively charged ion, is reversible. How could the pH be changed to ensure that virtually all the –OH groups have lost H^+?

At a suitable pH, a DNA fragment will have the same number of negative charges as it has nucleotides. Each extra nucleotide gives one extra unit of negative charge, but also increases the mass of the fragment. So all DNA fragments have almost exactly the same charge to mass ratio.

In gel electrophoresis, a mixture of DNA fragments is placed on a plate coated with jelly rather like that used on petri dishes to grow bacteria. An electrical voltage is applied across the ends of the gel. This draws the negatively charged fragments through the gel towards the positive electrode, see Figure 15.

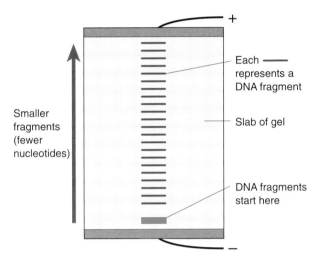

Figure 15 *Gel electrophoresis*   *Loading gel electrophoresis equipment*

Although each fragment has the same charge to mass ratio, the larger ones produce more drag as they move through the gel and thus move more slowly. So the fragments are separated by size as a series of bands and we can tell how many nucleotides each contains. More information about gel electrophoresis can be found in *More modern chemical techniques* (see *Further reading*).

Sequencing DNA fragments

The basis of the sequencing technique is to start with a single strand of DNA and build up a complementary stand from separate nucleotides using the enzyme DNA polymerase, which catalyses the reactions of one nucleotide with another that lead to DNA. Remember that if we know the sequence of the complementary strand of DNA, we can work out the sequence of the original by substituting each base with its complement – A for T, G for C *etc.* The trick is to add some specially made nucleotides that contain the sugar **dideoxyribose** (strictly 2,3-deoxyribose) rather than deoxyribose, Figure 16. Dideoxyribose has only one –OH group, so it can bond to only one phosphate group, rather than two.

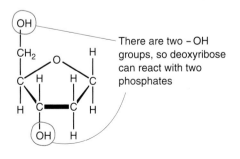

Figure 16 *Dideoxyribose*   *Deoxyribose*

So, if a nucleotide containing a dideoxyribose molecule is added to a growing DNA chain, no more nucleotides can be added, and the chain stops at that point. Dideoxynucleotides can be made containing each of the bases, C, G, A and T, so the chain can be stopped at any particular base and we will know which base is on the end of each fragment. In addition, these dideoxynucleotides have bonded to them a dye that fluoresces when light is shone on it (laser light is often used). C fluoresces blue, G yellow, T red and A green. This means that DNA fragments containing these dideoxynucleotides can be detected after electrophoresis by using laser light.

A simple example of this method is shown in the box *Sequencing – a simple example*. This technique is now automated and controlled by computers. It can be used to sequence DNA fragments up to 500 bases long. It allows laboratories such as the Wellcome Trust Sanger Institute, near Cambridge, to sequence up to 60 000 000 bases (60 Mb) per day. The Sanger Institute is named after Fred Sanger who developed the dideoxy technique for which he won one of his two Nobel Prizes.

Sequencing – a simple example

Imagine we have a single strand of DNA whose base sequence is TGCCAAGCT. (In a real example we would, of course not know this sequence.) We set up four test tubes each containing this strand of DNA, some DNA polymerase and a supply of nucleotides containing each of the four bases A, C, G and T. Tube 1 contains, in addition, a little dideoxyC. Tube 2 contains, in addition, a little dideoxyG. Tube 3 contains, in addition, a little dideoxyA. Tube 4 contains, in addition, a little dideoxyT. The situation is shown in Table 1.

Tube contents	1	2	3	4
Original DNA fragment	✓	✓	✓	✓
DNA polymerase enzyme	✓	✓	✓	✓
Normal nucleotides	A, C, G, T	A, C, G, T	A, C, G, T	A, C, G, T
Dideoxy nucleotide	DideoxyC	DideoxyG	DideoxyA	DideoxyT

Table 1 *The dideoxy method for sequencing DNA fragments*

In each tube, the formation of a complementary DNA sequence to the original will take place catalysed by the DNA polymerase enzyme. The complete complementary sequence is ACGGTTCGA. However, in Tube 1 the sequence will sometimes stop at the first C and sometimes at the second C because some of the dideoxyC will have been incorporated and this stops the chain growing. So Tube 1 will contain the following fragments: AC and ACGGTTC. Remember that there is only a little dideoxyC in the tube along with normal C molecules so not all the growing DNA chains will stop at the first C.

Similarly, in Tube 2 the sequence will sometimes stop at the first G, sometimes at the second G and sometimes at the third G because some of the dideoxyG will have been incorporated. So Tube 2 will contain the following fragments: ACG, ACGG and ACGGTTCG.

In the same way, Tube 3 will contain A and ACGGTTCGA.

In the same way, Tube 4 will contain ACGGT and ACGTT.

The short DNA fragments in each tube are then separated by gel electrophoresis, see Figure 17. The position of each fragment in the gel can be found by the colour of the fluorescent dye.

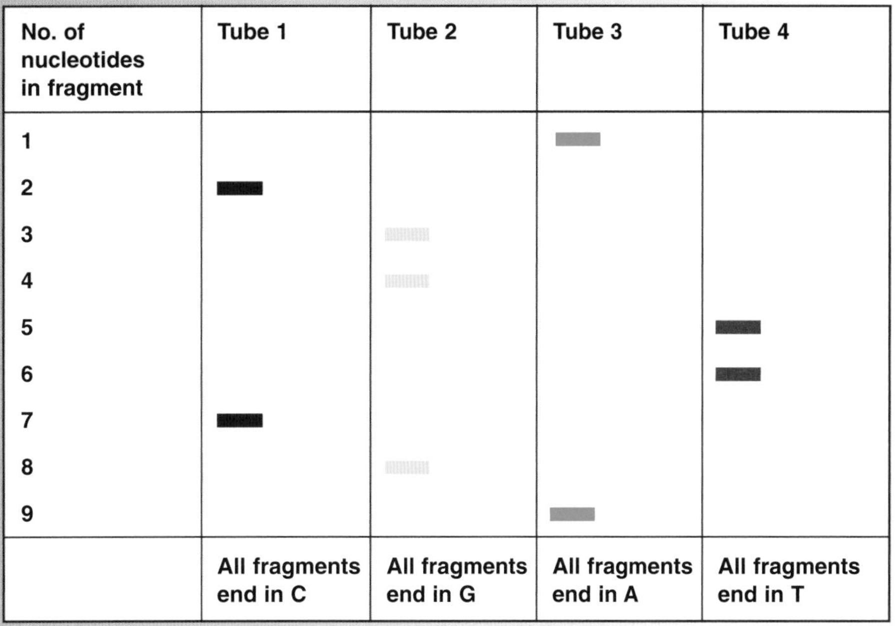

No. of nucleotides in fragment	Tube 1	Tube 2	Tube 3	Tube 4
1			▬	
2	▬			
3		▬		
4		▬		
5				▬
6				▬
7	▬			
8		▬		
9			▬	
	All fragments end in C	**All fragments end in G**	**All fragments end in A**	**All fragments end in T**

Figure 17 *The gel electophoretogram*

The result of the electrophoresis experiment (called an electrophoretogram) tells us that the first base in the complementary DNA chain must be A because there is a fragment in Tube 3 one base long and all fragments in Tube 3 end in A. The next base must be C because there is a two-base fragment from Tube 1 (which must end in C), so the sequence starts AC. The next base must be G because there is a three-base fragment in Tube 2 (which must end in G).

Continuing the analysis in this way leads to the sequence of the whole fragment as ACGGTTCGA. This is complementary to the original strand of DNA whose sequence must therefore have been TGCCAAGCT, see Figure 18.

No. of nucleotides in fragment	Tube 1	Tube 2	Tube 3	Tube 4
1			A	
2	A C			
3		A C G		
4		A C G G		
5				A C G G T
6				A C G G T T
7	A C G G T T C			
8		A C G G T T C G		
9			A C G G T T C G A	
	All fragments end in C	**All fragments end in G**	**All fragments end in A**	**All fragments end in T**

Figure 18 *Key to fragments in Figure 17*

Q 8. Figure 19 shows a diagram of the result of a sequencing experiment on a short section of DNA. Interpret the results and work out the sequence of bases on the original DNA.

No. of nucleotides in fragment	Tube 1	Tube 2	Tube 3	Tube 4
1		▬		
2	▬			
3	▬			
4			▬	
5				▬
6				▬
7			▬	
8		▬		
	This tube contains dideoxy C	This tube contains dideoxy G	This tube contains dideoxy A	This tube contains dideoxy T

Figure 19 *Gel electrophoretogram for Q8*

Fred Sanger

Figure 20 *Fred Sanger*

Fred Sanger, Figure 20, is a member of a very exclusive club indeed – those who have won two Nobel Prizes for science. His first prize was won in 1958 for determining the amino acid sequence of a protein, insulin – the first protein to be sequenced. This feat took Sanger and his Cambridge-based team ten years and involved the development of a number of new techniques including the use of 2,4-dinitrofluorobenzene (now called Sanger's reagent), a compound that attaches itself to amino acids. His method involved breaking a protein chain into fragments containing a few amino acids and separating these using two-way paper chromatography. The technique of protein sequencing is now automated and to sequence a simple protein such as insulin with 51 amino acids would be routine.

Twenty-two years later Sanger won his second prize, this time for sequencing DNA. He developed the dideoxy technique and the use of enzymes that cut a DNA chain at specific points.

The dideoxy method works well only for sections of DNA up to about 500 bases long. For sections larger than this, there are so many fragments on a gel plate that it becomes difficult to distinguish them.

Longer strands of DNA can be sequenced by first breaking them into lengths short enough to be sequenced by the dideoxy method. This can be done using enzymes that break the DNA chain at specific places (in a specific sequence of bases, for example). An alternative method, sonication (using ultrasound), breaks the DNA at random and is called 'shotgunning'. The short lengths of DNA will overlap each other and, after they have been separated, a computer can be used to look for overlapping sequences of bases on different strands and thus reassemble them in the right order. A simple example of this is shown in Figure 21.

A length of DNA was broken up at random and the resulting sections sequenced as follows:

```
G  C  A  A
T  C  G  G  G
C  G  G  G  C  C
C  T  A
C  A  A  T  C  G
```

By comparing the overlapping sequences, it is possible to reconstruct the original sequence as follows:

```
G  C  A  A
      C  A  A  T  C  G
                  T  C  G  G  G
                        C  G  G  G  C  C
                                    C  T  A
```

giving the original sequence as:

```
G  C  A  A  T  C  G  G  G  C  C  T  A
```

Figure 21 *Overlapping fragments of DNA*

Human DNA is not in one continuous length; it occurs as 23 pairs of chromosomes, Figure 22, each of which is a single double helix molecule. The DNA molecule in each chromosome is tightly coiled; a chromosome is about 0.6 µm in diameter but if the DNA it contained were unwound, it would be several centimetres long. Parts of each chromosome form **genes** – lengths of DNA that contain the information to determine a single characteristic such as blue or brown eye colour. They do this by holding the 'recipe' to make a single protein. The whole genome contains about three billion (3×10^9) nucleotides but only about 3% of this makes up actual genes, the function of the rest is unknown and it is often called **'junk' DNA**.

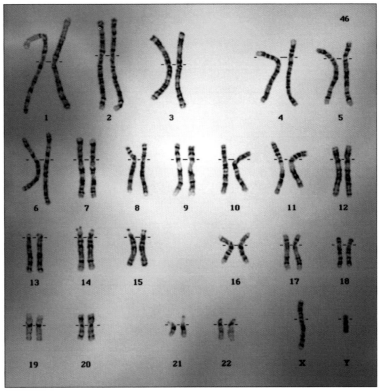

Figure 22 *23 pairs of chromosomes in human DNA*

Well before DNA sequencing became possible, scientists were able to produce 'maps' that showed where particular genes were situated, *ie* on which chromosome and whereabouts on that chromosome. Locating a gene on a chromosome is done by tracing how characteristics are inherited within families. Genes close together on the same chromosome will tend to be inherited together. An example of this the inheritance of a condition called nail-patella syndrome (NP), in which sufferers have small, discoloured nails and missing kneecaps, and A or O blood groups. Almost all sufferers of NP have blood groups A or O, not B or AB.

Later, maps were produced by staining chromosomes with a dye that binds to the base guanine, G. These showed how the frequency of this base varied along the chain of DNA. This technique allows us to identify areas rich in G and those where G is less abundant and therefore provides some points of reference. The next step was to use short lengths of DNA as 'probes'. These probes bind to the DNA at sites where they meet a complementary sequence of bases (so a probe with the sequence ACGGTA would bind to the sequence TGCCAT, for example).

These sites are called markers. Another technique for gene mapping is the use of enzymes called restriction enzymes that will cut a strand of DNA at a particular sequence of bases. For example the enzyme known as *EcoRI* will make a cut in a DNA molecule between the G and the A of the sequence GAATTC. This results in fragments of DNA with known sequences of bases at each end. These can be separated by gel electrophoresis.

These techniques allow researchers to produce a map of the human genome that locates particular genes and that has markers to identify particular regions, see Figure 23. Helped by this map, researchers can then sequence sections of DNA knowing where they come from.

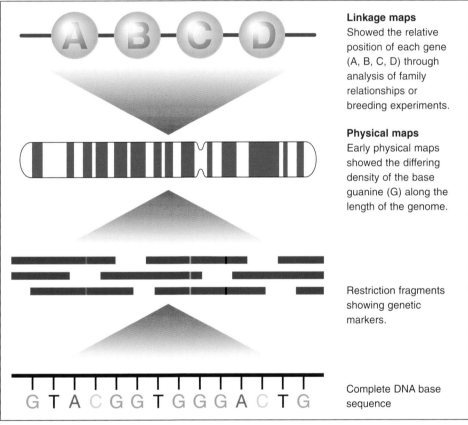

Linkage maps
Showed the relative position of each gene (A, B, C, D) through analysis of family relationships or breeding experiments.

Physical maps
Early physical maps showed the differing density of the base guanine (G) along the length of the genome.

Restriction fragments showing genetic markers.

Complete DNA base sequence

Figure 23 *Different types of genetic maps*

Different approaches to genome sequencing

We saw in the box *Private versus public – different approaches to the genome*, that there are two organisations, The International Human Genome Sequencing Consortium (a group of scientists at 16 institutions around the world, often abbreviated to HGP) and Celera, working on the human genome. In addition to the moral debate between HGP and Celera over patenting, there is also an interesting difference in the methods they use. HGP works from a gene map, taking sections of DNA from particular areas of the genome and then sequencing the bases. Celera uses what has been called the 'whole genome shotgun' (WGS) approach. This involves breaking the whole genome into random lengths, sequencing them and then using powerful computers to reassemble the information on the basis of overlapping sequences of bases. The second approach is potentially faster because it misses out the mapping phase. However, it does require a great deal of computing power because the whole genome contains very many repeated sequences of bases that could easily be wrongly placed. Indeed the WGS approach does not appear to work particularly well for the human genome.

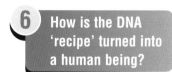

6 How is the DNA 'recipe' turned into a human being?

The answer to this question involves proteins. Proteins make up much of the human body and, when behaving as enzymes, control the reactions that make all the other complex molecules in it. The information held by DNA molecules is essentially a set of instructions to make a range of proteins.

Proteins are poly-amino acids. There are thousands of different proteins with a vast range of properties. The properties of a protein are governed by its shape, which in turn is governed by its sequence of amino acids. The sequence of bases on a stretch of DNA 'tells' other molecules (called RNA) the correct order to assemble amino acids to make proteins. Each sequence of three bases identifies, or 'codes for', a particular amino acid and is called a **codon**.

There are four bases in DNA and this means that there are 64 (= 4^3) ways of arranging them in threes. (This is an example of a general formula that the number of arrangements is x^n where x is the number of items and n the number of them in the group. You might like to check that this formula works by writing out all the ways of arranging A, C, G and T in pairs – you should find $4^2 = 16$.) Sixty-four is more than enough combinations to code for the 20 amino acids that are found naturally in proteins. So there is more than one three-letter code for each amino acid. There are also codons that signify the beginning and the end of a protein chain. One way of summarising the code is shown in Figure 24. Reading from the inside out gives the amino acid represented by a particular three-letter sequence of bases in a length of DNA. This code is practically the same in all living creatures.

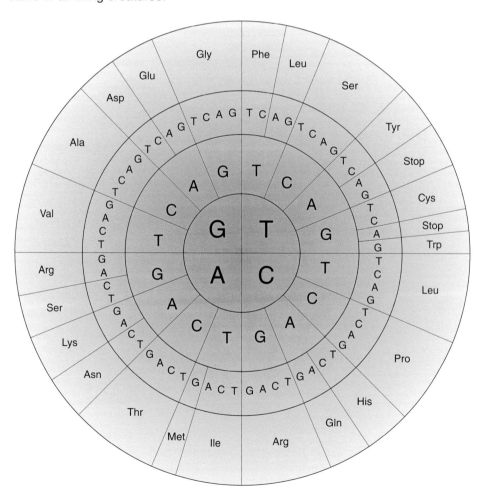

Figure 24
The genetic code. This figure shows how the three letter codons of DNA relate to amino acids. Working from the inside of the circles gives a three letter code and the corresponding amino acid. For example, GGA codes for glycine (Gly). The codon TAA ends a chain of amino acids

Q 9. Why would a code based on two-letter codons not work for assembling proteins from the 20 naturally-occurring amino acids?

How do we know what codon refers to what amino acid?

The first 'word' or codon of the genetic code was 'read' in 1961 when two researchers, Johann Matthae and Marshall Nirenberg made a length of RNA that contained only the base uracil (U), see below and Figure 29, so the only codon it contained was UUU. They placed this in a solution containing a mixture of all 20 naturally occurring amino acids and found that a protein was made that contained only the amino acid phenylalanine repeated over and over again. Thus the codon UUU represented phenylalanine.

The principle of the process by which amino acids are assembled in the right order to make a protein as specified by a section of DNA is straightforward even if the details are complicated. The process depends on polymers called **RNA (ribonucleic acid)**. These molecules are constructed from nucleotide monomers. These are similar to DNA molecules but with two differences:

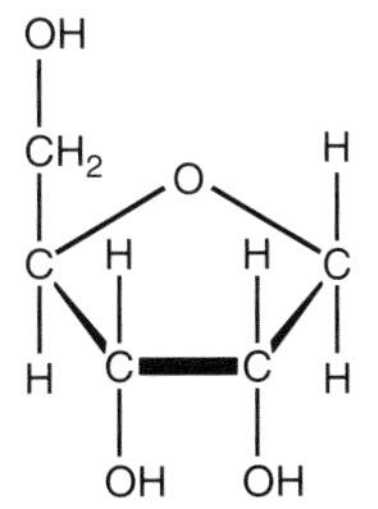

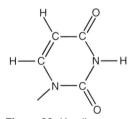

Figure 25 *Ribose*

1. The sugar part of the molecule is **ribose**, rather than deoxyribose. Ribose has an extra oxygen atom compared with deoxyribose, see Figure 25.

2. In RNA, the base uracil (U) occurs in place of thymine (T). Uracil has a hydrogen atom where thymine has a –CH_3 group, but otherwise the molecules have the same shape, see Figure 26. In particular, uracil can hydrogen bond with adenine in just the same way as thymine can.

Figure 26 *Uracil*

Q 10. Look at the structure of the ribose molecule in Figure 25 and compare it with that of deoxyribose in Figure 3. Identify the extra oxygen atom. What group is it a part of?

Q 11. Draw diagrams to show a) thymine and b) uracil hydrogen bonding with adenine. Explain why the extra –CH_3 group in thymine makes no difference to this bonding.

A molecule of RNA can store information in its sequence of bases in just the same way that DNA can.

In brief, the protein synthesis takes place as follows.

A section of DNA in the nucleus of a cell 'unzips' in the same sort of way as in replication. In this case, controlled by different enzymes, molecules of free ribose-containing nucleotide monomers of the correct type pair up with the newly exposed bases on the two unravelled strands (A pairing with U in this case). These monomers, held in place by hydrogen bonding then link together by phosphate-sugar bonding as in replication. The original strand of DNA acts as a template for a new molecule of RNA. This new molecule is called **messenger RNA** or **mRNA** for short. The process is similar to the replication of DNA, but the resulting molecule of mRNA is a single strand rather than a double helix. mRNA contains the same genetic information as the original DNA that formed the template for it. The mRNA passes through pores in the nucleus of the cell and into the **cytoplasm** where it encounters bodies called **ribosomes**, made of RNA and protein, see Figure 27. This is where the proteins are actually made from individual amino acids that are found in the cytoplasm, using molecules called **transfer RNA** (**tRNA** for short). Transfer RNA is a sort of adaptor molecule, it translates the 'language' of DNA and RNA (a sequence of base) into the 'language' of proteins (a sequence of amino acids). That is it translates each three-base codon on the mRNA molecule into the correct amino acid. It consists of molecules of RNA with three letter base sequences on one end, see Figure 28. The other end of the molecule can bond to an amino acid. tRNA molecules with different base sequences bond to different amino acids as shown in Figure 29.

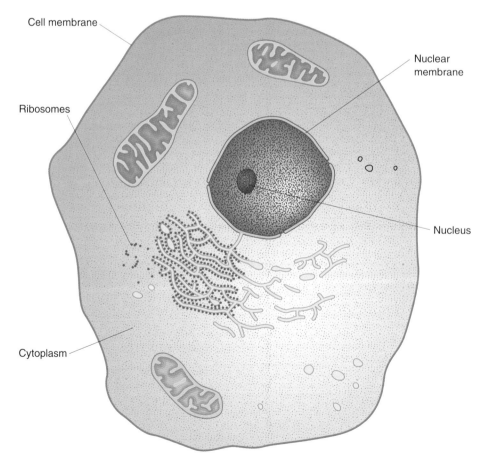

Figure 27 *An animal cell*

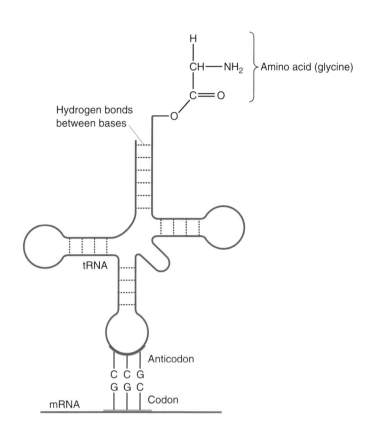

Figure 28
A schematic diagram of a tRNA molecule. The continuous line represents a length of RNA in a characteristic clover leaf shape. The dotted lines represent hydrogen bonds between bases. The important feature of the molecule is that on one end it has an 'anticodon', the three complementary bases to the codon on mRNA, and on the other end has the appropriate amino acid. So the codon GCC has the anticodon CGG and this represents alanine.

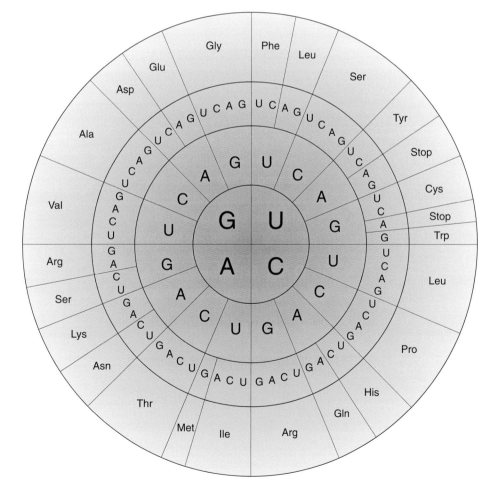

Figure 29
The genetic code. This figure shows how the three letter codons of mRNA relate to amino acids. Working from the inside of the circles gives a three letter code and the corresponding amino acid. So, for example, GCA codes for alanine (Ala). The codon UGA ends a chain of amino acids.

Q 12. How does figure 29 relate to figure 24?

On the ribosome, a molecule of tRNA hydrogen-bonds to a codon (a three-base section) of the mRNA molecule *via* its three bases (called an **anti-codon**). A second tRNA molecule bonds to the next three-base codon in the same way. These two tRNA molecules hold two amino acids close to one another so that they can bond together and form a dipeptide. The ribosome now moves three bases along the strand of mRNA. A third tRNA molecule now bonds to the next three bases. Its amino acid joins the growing chain and the first tRNA molecule (minus its amino acid) is released. This process continues and the amino acid chain lengthens to form a polypeptide and then a protein. This process joins the amino acids in the correct sequence to make a particular protein, see Figure 30.

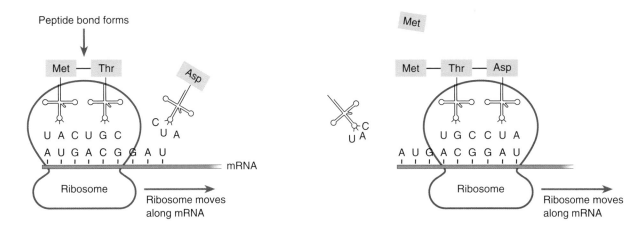

Figure 30 *Protein synthesis*
Two tRNA molecules hydrogen bonded to the mRNA molecule via their anti codons. A peptide bond forms linking the two amino acids.

A third tRNA molecule bonds to the mRNA as the first one is released. (It will later bond to another free amino acid). The third amino acid adds to the growing chain.

This sequence of amino acids is called the **primary structure** of the protein. A 'stop' codon signals the last amino acid to be added and the protein molecule is released. When the protein is released from the tRNA, its amino acid chain coils up and the coil then folds to make a unique shape. It is this shape that governs the properties of the protein. The coils and folds are held in place by hydrogen bonding (and other intermolecular interactions).

Q 13. What molecule is eliminated when two amino acids join to form a dipeptide? What type of reaction is this?

Q 14. The sequence of amino acids in a protein is called its primary structure. What terms are used for a) the coiling of a protein and b) the folding of an already-coiled protein?

Q 15. The structures of the amino acids glycine (gly) and alanine (ala) are given in Figure 31. Draw the dipeptide gly-ala and mark the groups that could hydrogen bond to other groups in a protein chain.

Figure 31 *The structures of glycine (left) and alanine (right)*

To sum up, although the details of protein synthesis from the information held in a DNA molecule are complex (and the fine detail more complex still), the principle is fairly simple. The actual instructions for making a protein (by assembling amino acids in a particular order) are held in sections of the DNA molecule by the sequence of bases. Each three base sequence (called a codon) relates to a single amino acid. The information can be copied by the process of the DNA double helix 'unzipping' and a new strand of DNA or mRNA building itself round the original. This copying process depends on hydrogen bonding in two ways. Firstly each base can only hydrogen bond effectively to its complementary base (G to C and A to T or U). Secondly, hydrogen bonds can break and reform under conditions in which covalent bonds stay intact, so the original DNA helix is unaffected. tRNA acts as an 'adaptor' molecule with bases on one end and an amino acid on the other. mRNA acts as a sort of intermediary between DNA and tRNA, taking the information from the genes in the nucleus of the cell to the site of protein synthesis (the ribosomes) in the cytoplasm.

A section of a DNA molecule that codes for one protein is called a **gene**. We now know that the human genome has around 30 000 genes. Not all our DNA, however, codes for proteins. Large sections of it (up to 97%) do not code for proteins and are sometimes called '**junk DNA**'. No one is sure why this is – it may be redundant copies of once-useful DNA. One theory is that there may be an analogy with a computer's hard disk. On most people's hard discs there are many files that are no longer required – they may be parts of projects that are now finished or there may be several earlier drafts of documents. Junk DNA may be something like this – once-useful sequences that are now not needed, or multiple copies. When you back up your hard disc, you might simply copy its whole contents (including all the unwanted files and drafts) rather than just the useful information. However, no-one is certain about the origin of junk DNA, and research is continuing.

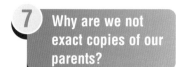

7 **Why are we not exact copies of our parents?**

Recombination of genetic material

Although they resemble both their parents, children are not exact copies of either. This is due to the differences between the way in which cells reproduce and sexual reproduction.

Each cell in the human body contains 23 pairs of chromosomes that contain the DNA. When cells divide during growth, for example, the chromosomes are copied to give double the number, half of which pass into one of the two new cells and half into the other. Provided there have been no mistakes in copying, the two new cells formed by division have identical chromosomes to the original cell.

However, sexual reproduction is different – it occurs via sex cells called **gametes** (sperm and eggs) that have only a single set of 23 chromosomes. To make sex cells, the 46 chromosomes of an ordinary cell (23 inherited from the father and 23 from the mother) first pair up. A process called **recombination** occurs, see Figure 32. Matching bits of DNA from the two chromosomes swap over. For example a gene for blue eyes (strictly an allele of the gene) from the mother might swap with a gene for brown eyes (strictly an allele of the gene) from the father. This makes two new chromosomes each with a mixture of genetic material from each of the two parents. The cell then divides in the normal way, the new chromosomes being copied, to give two new cells each with a set of 46 chromosomes (23 pairs). A second cell division then takes place *without* chromosome copying to give four new cells with just 23 chromosomes each. These are the sex cells – in males they become sperm and in females they become eggs. When a sperm from a father fertilises an egg from a mother, a cell with 46 chromosomes (23 from the father and 23 from the mother) is formed. This then divides successively in the normal way and develops eventually into a child.

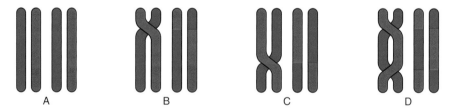

Figure 32 *During recombination DNA is swapped between two paired chromosomes*

So every cell in an individual contains 23 pairs of chromosomes that are identical to the chromosomes in every other cell in that individual. One pair of chromosomes contains a mixture of genetic information from its paternal grandparents (the mother and father of its father). The second pair of chromosomes contains a mixture of genetic information from its maternal grandparents (the mother and father of its mother).

This mixing of genetic information is important for evolution – it means that a species can continually 'try out' new combinations of genes. Those that are successful will produce individuals that thrive and pass on their genes to their offspring. Individuals with less successful combinations of genes may die before reproducing and may also be less successful in attracting mates – both of which mean that their genes will not be passed on.

Copying errors

From time to time there will be errors in copying DNA – perhaps the wrong nucleotide will be copied. This may result in a wrong amino acid being placed in a protein. In many cases this will cause no harm to the individual concerned but in other cases it may cause a problem such as a disease. The condition sickle cell anaemia is caused by a single incorrect amino acid caused by a single incorrect base. Occasionally these sorts of errors will confer some advantage on the individual concerned. Such individuals will thrive and be more likely to pass on their genes to their offspring and they to their offspring in turn. Conversely, where changes in DNA cause a disadvantage to the individual, the individual will be less likely to thrive and pass on its genes. This basic principle, that the fittest survive and pass on their genes while the less fit do not, is what drives evolution.

Q16. Some changes to base sequences will cause no change to the proteins produced. Look at Figure 24 and try to explain how this might come about.

Mutations

Occasionally, DNA can be chemically changed in some way, possibly by radiation that can break chemical bonds or by exposure to some chemical. This too contributes to the process of evolution by producing altered base sequences and thus altered genes. Most of these will be deleterious changes and may cause the organism to die before passing on its genes to its offspring. Occasionally the new DNA will be advantageous to the organism and will be passed on into the pool of genes of the species.

8 **The future of genome research**

It would be easy to think that, with the human genome almost completely sequenced, there is no future for genetic research. However, this would be far from the truth. Much remains to be done on the human genome itself and work has barely started on a host of applications that promise to transform healthcare in the next few years.

Completing the genome sequence

What has been published so far is merely a draft of the complete genome, amounting to about 90% of it – a final version is not expected until 2003. Furthermore, what has been done so far is to produce the sequence of bases only. Much of this sequence has not yet been organised into genes. This means that it is not always known where one gene finishes and another starts and which DNA codes for proteins and which is 'junk'.

The human proteome

As we have seen, genes code for proteins. Some proteins cause genetic diseases if cells produce the wrong protein, or produce too little or too much of the right protein. So the next challenge for chemists is to identify and understand the range of proteins in the human body – the human **proteome**. This will be more difficult than understanding the genome because the current techniques for identifying and sequencing proteins (such as 2-D gel electrophoresis) are less efficient than those used for sequencing DNA.

One method of finding which proteins are important in diseases is to compare the proteins found in healthy tissue with those in diseased tissue. Once these proteins are identified and their structures known, it may be possible to design a drug molecule that will interact with them and therefore affect the course of the disease.

Gene chips and diagnosis

A gene chip consists of strands of DNA fixed to a grid made of glass. If a DNA-containing sample taken from a patient is passed over the chip, the immobilised DNA strands on the chip will bind to any complementary DNA sequences in the sample. This technique may enable us to find out whether the sample comes from a virus or from a bacterium, for example. This is done by tailoring the DNA strand on the chip to pick out complementary strands of DNA that are specific to a virus or to a bacterium. Knowing whether a disease is caused by a virus or a bacterium will help in diagnosis. For example, the first symptoms of flu (a relatively minor disease in most patients) and of meningitis (potentially fatal in most cases) are very similar but the two diseases would be easily distinguished on the basis of a genetic test. Flu is caused by a virus, and meningitis by a bacterium.

Personalised treatment

There is only about a 0.1% difference in the genomes of different humans. However, this will cause different individuals to react differently to the same drug treatment. Gene chip techniques offer the possibility that drug treatment (type and dose of drug) may be able to be tailored to individuals.

Antibodies and antisense treatments

Our immune system recognises and combats disease-causing 'foreign' molecules, such as certain proteins, and produces antibodies that bond specifically to them and neutralise their effects. Extra supplies of antibodies to help our bodies combat specific proteins identified by research on the genome might be able to be made.

Antisense treatments act even closer to the heart of the problem by targeting not the foreign protein but on the section of DNA that codes for it. They consist of strands of DNA that are complementary to the mRNA that is involved in making the protein. They therefore bond to the mRNA and prevent it making the disease-causing protein.

Gene therapy

Gene therapy for a disease involves changing the genetic code (the DNA) in certain cells (or the whole organism) so that the body's cells themselves produce a protein that is needed to combat a disease. Some trials are underway, but 20 – 30 years may be the sort of timescale before this sort of treatment is in regular use.

So, although the sequencing of the human genome is close to being completed, there are plenty of challenges and opportunities for chemical scientists working in this area in the future.

9 Further reading / websites

Books

F. Crick, *What mad pursuit*, London: Penguin, 1990.

R. Dawkins, *The blind watchmaker*, London: Penguin, 1988.

R. Dawkins, *The selfish gene*, Oxford: Oxford University Press, 1989.

J. Gribin, *In search of the double helix*, London: Corgi, 1985.

S. Jones, *The language of genes*, London: Flamingo, 1994.

R. Levinson, *More modern chemical techniques*, London: Royal Society of Chemistry, 2001.

M. Ridley, *Genome: The autobiography of a species in 23 chapters*, London: Fourth Estate, 1999.

J. D. Watson, *The Double Helix,* Harmondsworth: Penguin Books, 1970.

E. J. Wood, C. J. Smith and W. R. Pickering, *Life Chemistry and molecular biology*, London: Portland Press, 1997.

Articles

J. Evans, *Chemistry in Britain*, **2001, 37 (8)**, 26.

Websites

Celera Genomics: www.celera.com (accessed May 2002).

The Sanger Centre: www.sanger.ac.uk/ (accessed May 2002).

The Wellcome Trust: www.wellcome.ac.uk (accessed May 2002).

UK Human Genome Project: www.hgmp.mrc.ac.uk (accessed May 2002).

Glossary

The glossary contains short explanations of the words in bold in the text.

Anti-codon

A sequence of three nucleotide bases that binds via hydrogen bonding to a codon. Each base binds to its complementary base so the codon CCG has the anti-codon GGC.

Base

In general, a species that can accept a proton. In this context, one of the five nitrogen-containing organic molecules adenine, cytosine, guanine, thymine and uracil. See Figures 4 and 26.

Base pairing

The formation of hydrogen bonds between the bases C and G and between A and T that holds together the DNA double helix, see Figure 8.

Chain length

The number of monomers in a polymer chain.

Chromosome

One of a group of thread-like structures found in the nuclei of cells, containing protein and DNA in roughly equal amounts. Chromosomes contain the genetic information and are duplicated when cells divide.

Codon

A sequence of three nucleotide bases that corresponds to a particular amino acid. For example the codon GGC codes for glycine.

Complementary

Complementary bases are those that hydrogen bond effectively to each other – C and G, and A and T (or U).

Cytoplasm

The jelly-like material surrounding the cell nucleus where most of the reactions of the cell take place.

Deoxyribose (3-deoxyribose)

A sugar molecule, similar to ribose but with one fewer oxygen atom, found in DNA, see Figure 3.

Dideoxyribose (2,3-deoxyribose)

A sugar molecule, similar to ribose but with two fewer oxygen atoms, see Figure 16. It is used in sequencing DNA and is not naturally-occuring.

Dideoxy method

A method of sequencing DNA fragments in which growing chains of DNA are stopped by the incorporation of a nucleotide containing dideoxyribose.

DNA, deoxyribonucleic acid

A polymer molecule, found in chromosomes, that carries genetic information. It contains the sugar deoxyribose.

Electronegative

Electronegative atoms (nitrogen, oxygen and fluorine) have the ability to attract the electrons in covalent bonds towards them.

Gametes

Reproductive cells – they contain half the number of chromosomes contained in normal cells.

Gel electrophoresis

A technique used for separating mixtures of charged molecules according to their size, by using an electric field to drag them through a gel.

Gene

The unit of heredity composed of DNA. One gene contains the information to make a single protein.

Genome

All the genes in a single set of chromosomes from one organism.

Hydrogen bonds

Intermolecular forces that act between a hydrogen atom, covalently bonded to a nitrogen, oxygen or fluorine atom, and another nitrogen, oxygen or fluorine atom.

Junk DNA

DNA that does not code for proteins. Its function is unknown but it makes up about 97% of the human genome.

Messenger RNA (mRNA)

A type of RNA that carries genetic information held by DNA from the nucleus of the cell into the cytoplasm.

Nucleoside

A molecule consisting of a sugar molecule linked to an organic base. It is like a nucleotide (see below) but without the phosphate group.

Nucleotide

A molecule consisting of a sugar, a phosphate group and an organic base, see Figure 3. It is the monomer from which DNA is built.

Peer review

The process by which scientists read and approve for publication papers (articles detailing their research) by their colleagues.

Primary structure

The sequence of amino acids that makes up a polypeptide.

Phosphate

A group of atoms based on salts of phosphoric acid, H_3PO_4, see Figure 3.

Proteome

The range of proteins found in an organism.

Recombination

The regrouping of genes that occurs during sexual reproduction.

Replication

The copying of DNA.

Ribose

The sugar molecule of RNA, see Figure 25.

Ribosome

A structure made up of protein and DNA found in the cell: the site where protein synthesis takes place. There are many thousands in a typical cell.

RNA, ribonucleic acid

A polymer molecule, found in cells. One type, messenger RNA, translates the genetic information held by DNA into proteins. It contains the sugar ribose.

Sonication

The random splitting up of DNA into fragments, brought about by ultrasound.

Transfer RNA (tRNA)

A type of RNA that brings amino acids to the ribosomes so that they can be assembled into proteins according to instructions held originally by DNA molecules.

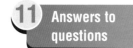

1. The nitrogen atoms have lone pairs of electrons that are capable of accepting a H+ ion.

2. The DNA molecule has many phosphate groups. Each of these has a free –OH group which is capable of releasing a H+ ion leaving behind an –O− group.

3. A water molecule. A condensation (or elimination) reaction.

4. TCCAGTTA.

5. Inspection of models. Students should see that A and T can form two hydrogen bonds and G and C can form three.

6. The C–C bond energy is 347 kJ mol⁻¹ and an H ‑‑‑ O–H hydrogen bond is about 25 kJ mol⁻¹. Students should appreciate that a hydrogen bond is more than ten times weaker than a covalent bond but that the overall effect of many hydrogen bonds can hold molecules together quite tightly.

7. Le Chatelier's principle predicts that reducing the concentration of H+ ions will force the –OH groups to dissociate, so a high pH should be used.

8. The complementary sequence is GCCATTAG, see Figure 33. Therefore the original sequence was CGGTAATC.

No. of nucleotides in fragment	Tube 1	Tube 2	Tube 3	Tube 4
1		G		
2	GC			
3	GCC			
4			GCCA	
5				GCCAT
6				GCCATT
7			GCCATTA	
8		GCCATTAG		
	This tube contains dideoxy C	This tube contains dideoxy G	This tube contains dideoxy A	This tube contains dideoxy T

Figure 33 *Electrophoretogram*

9. There would only be 4² =16 combinations – not enough for one for each of the 20 naturally occurring amino acids.

10. The oxygen in question is part of an alcohol (–OH) group.

11. See Figure 34.

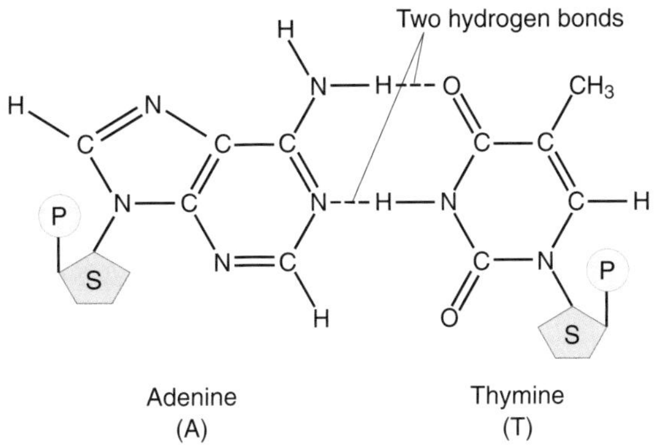

Figure 34a *Thymine hydrogen bonding with adenine*

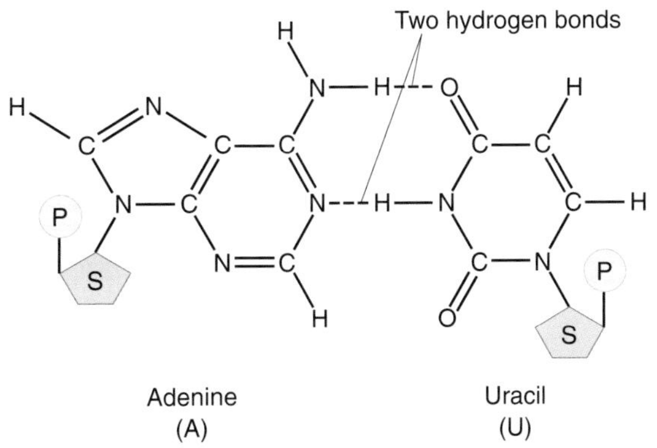

Figure 34b *Uracil hydrogen bonding with adenine*

The –CH$_3$ group is not involved in the hydrogen bonding. Hydrogen bonding takes place only between a hydrogen atom covalently bonded to an electronegative atom (nitrogen, oxygen or fluorine) and another electronegative atom.

12. Figure 29 has uracil molecules in place of thymine in Figure 24.

13. Water. A condensation reaction.

14. Coiling is the secondary structure, and folding the tertiary structure.

15. See Figure 35

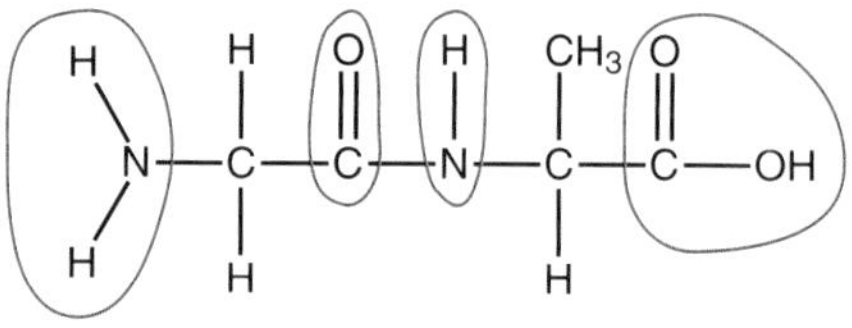

Hydrogen bonding groups are ringed

Figure 35 *The gly-ala peptide*

16. Most amino acids have more than one codon that codes for them. A change of one base in a codon could still result in a codon for the same amino acid. For example AGA and AGG both code for arginine.

12 **Teachers' notes**

This section amplifies some of the ideas in the main text for the use of the teacher. Teachers will use their own discretion as to whether or not to discuss these points with students.

X-ray diffraction

X-ray diffraction still cannot be used to determine the base sequences in DNA because these do not form a regular repeating pattern.

Hydrogen bonding

At room temperature, individual hydrogen bonds will break and reform. If DNA is heated to between 70 °C and 90 °C, hydrogen bonds in bulk will break and the two helices will separate. This occurs over a range of temperatures depending on the bases involved – it is harder to break the three hydrogen bonds between G and C than the two hydrogen bonds between A and T.

Polypeptides and proteins

The difference between a polypeptide and a protein is a moot point. Some authorities draw a distinction based on the number of amino acids, with the dividing line being of the order of 50 amino acids – shorter than this is considered to be a polypeptide and longer a protein. An alternative distinction is to consider a sequence of amino acids to be a polypeptide but when coiled and/or folded it is regarded as a protein.

Representation of phosphate, sugar and base units

These have been represented in most structure diagrams as:

phosphate – a yellow circle

sugar – a purple pentagon

base – a green hexagon (or fused hexagon and pentagon)

respectively. This allows students to concentrate on and appreciate the overall shapes of complex molecules without becoming bogged down in the detail of their full chemical structures.

However, in detail these representations must be interpreted with some caution. In particular, when a sugar-phosphate link forms, a molecule of water is eliminated. This means that the yellow circle representing phosphate in a free nucleotide does not represent exactly the same group of atoms as it does in a section of DNA as shown on p.46.

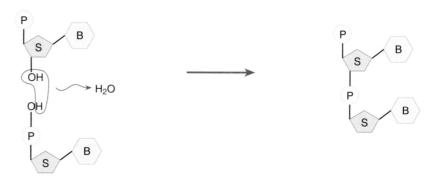

Teachers should be aware of this point, but to what extent they point it out to students is left to their discretion.